A Cartoon Guide to Becoming a Doctor

by
Dr. Fizzy McFizz

A Cartoon Guide to Becoming a Doctor

© 2011 Fizzy McFizz. All rights reserved.

ISBN 978-1-105-09102-5

This book is a work of fiction. The names, characters, incidents and places are the products of the authors' imagination, and are not to be construed as real. None of the characters in the book is based on an actual person. Any resemblance to persons living or dead is entirely coincidental and unintentional.

Table of Contents

1. Pre-med..5

2. Medical School 9

3. Residency 57

4. Attendings 85

5. Haikus.. 95

6. The Specialties... 99

7. Everything Else..105

1

PRE-MED

SHOULD YOU GO TO MEDICAL SCHOOL??

YOUR MED SCHOOL INTERVIEW

4 SIGNS YOU MIGHT BE PRE-MED

YOU'RE VOLUNTEERING FOR... SOMETHING

YOU DREAM OF "ORGO"

YOU'VE CRIED OVER A GRADE

YOU MAKE HYSTERICAL POSTS ON PREMED MESSAGE BOARDS
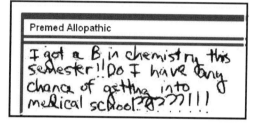

2

MEDICAL SCHOOL

THE 32 TYPES OF MED STUDENTS

SMARMY	PRETTY ONE	FUTURE ORTHOPOD	NERD	CREEPY NERD
GRANDPA	FOREIGN GUY	FUTURE OB/GYN	VOLUNTEER	GIRL NERD
GUNNER	SPOILED PRINCESS	FORMER EMT	CUTIE PIE	HEARTTHROB
RELIGIOUS KID	FORMER I-BANKER	MOMMY	PLAYBOY	FEMINAZI	SLUT
YOUNGSTER	YOUNGSTER'S BABY SISTER	FUTURE PSYCHIATRIST	DUMB ONE	MD/PhD
NEVER SPEAKS	WON'T SHUT-UP	CHEATER	FINK	DOWNER	OVERCAFFEINATED

GUARANTEES FOR YOUR FIRST WEEK OF MEDICAL SCHOOL:

YOU WILL PURCHASE HIGHLIGHTERS

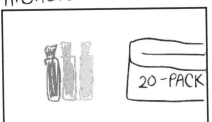

YOU WILL GET AWFUL EYESTRAIN

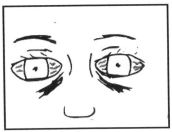

YOU WILL SMELL

YOU WILL NOT LEARN ABOUT ANY ACTUAL DISEASES OR ANYTHING USEFUL

A RELATIVE WILL ASK YOU A MEDICAL QUESTION

YOU WILL SLIP NEW WORDS INTO CASUAL CONVERSATION

YOU WILL FEEL "GUILTY" EVERY SECOND YOU AREN'T STUDYING

Most Common Questions Asked by First Year Medical Students

Why does my cadaver smell so bad?

Will the Krebs cycle ever be useful to me? (Answer: No)

Why do I keep getting alcohol poisoning?

Ways to Procrastinate from Studying

Count hairs on head

Check email every 15 seconds

High-fat fast food run

Sing "Old Time Rock 'n Roll" in underwear

Farmville

Dress up bunch of monkeys, have them reenact Civil War

HOW TO MAKE YOUR MED SCHOOL RELATIONSHIP WORK:

ALLOW YOUR SIGNIFICANT OTHER TO START THE DATE WITHOUT YOU

HAVE YOUR SO QUIZ YOU ON THE CRANIAL NERVES AND CALL THAT A DATE

IN A PINCH, YOUR ANATOMY CADAVER CAN ACT AS A STAND-IN ON YOUR DATE.

TERMS YOU CAN NO LONGER LAUGH AT ONCE YOU BECOME A DOCTOR

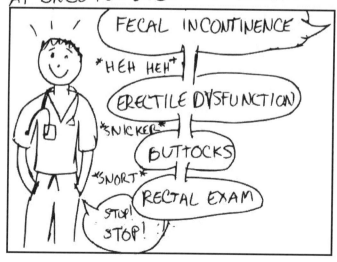

MOST COMMON VALENTINE'S DAY PRESENTS GIVEN BY MED STUDENTS:

HEART FROM CADAVER

FOLEY BAG

A FREE COFFEE AT STARBUCKS

THE 6 TYPES OF MED SCHOOL COUPLES

LOVEY DOVEY

COMPETITORS

TUTOR/TUTEE

SECRET ROMANCE

STUDY BUDDIES

DRAMA KING & QUEEN

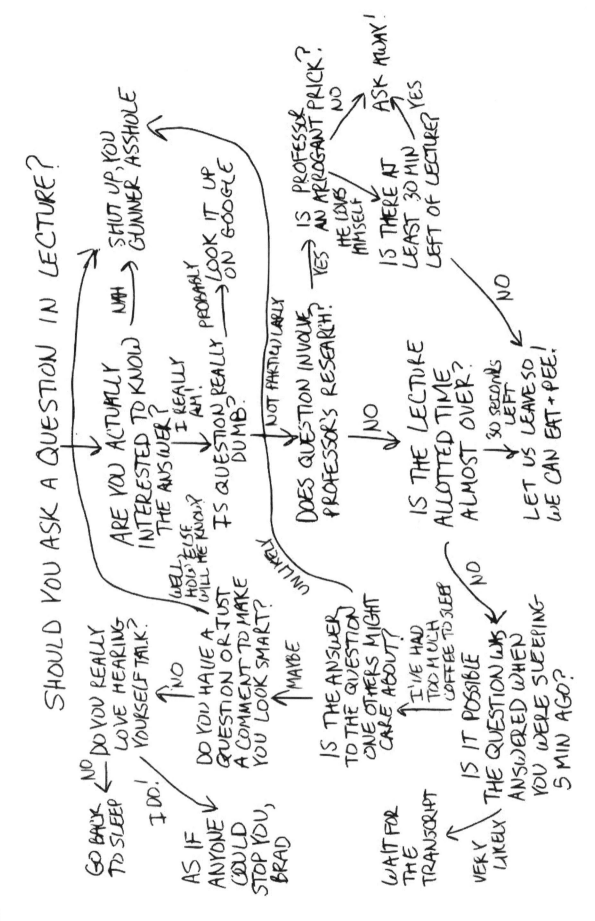

STAGES OF STAYING UP ALL NIGHT STUDYING FOR AN EXAM

1: ENTHUSIASTIC

2: ORGANIZED

3: OVERWHELMED

4: ANGRY

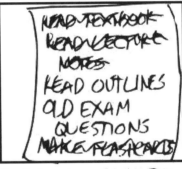
5: RE-ORGANIZED

6: VENDING MACHINE

7: 5 MINUTE PEE

8: PANIC

9: CONTEMPLATING ALTERNATE CAREERS

10: SURREAL

11: APATHY

12: EXAM

Happy Anti-Valentine's Day!

When I was in my first year of med school, a couple of my friends and I decided that we were sick of Valentine's Day and how it made single people feel crappy about themselves. So we decided to celebrate Anti-Valentine's Day. This holiday would express feelings that weren't specifically of love, but weren't of hate either.

We went out to the back of the university hospital and we climbed through a dumpster and selected "presents" for all our friends. We then taped little messages on each of the presents that expressed the sentiments of Anti-Valentine's Day. I can't remember, but I'm assuming we were drunk at the time.

Here were a few examples of presents:

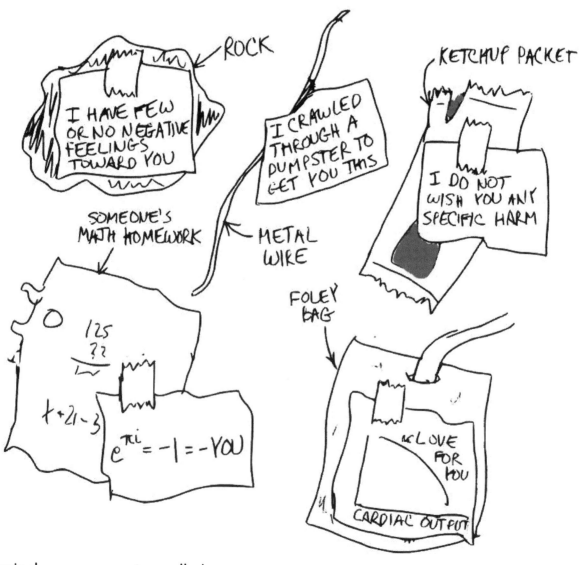

Amazingly, we were not expelled.

PERSONALITY DISORDERS COMMONLY SEEN IN MED STUDENTS*

OBSESSIVE-COMPULSIVE

SCHIZOID

NARCISSISTIC

PARANOID

HISTRIONIC

SCHIZOTYPAL

*WILL ONLY GET WORSE AS TRAINING PROGRESSES

Ways Med School is Like Junior High

THE 8 TYPES OF MED SCHOOL PROFESSORS

THE ENTHUSIAST

ADVANTAGES: WILL DO YOUR DISSECTION FOR YOU.
WARNING: ANATOMY IS NOT FUN

THE DRONE

ADVANTAGES: CATCH UP ON SLEEP DURING CLASS
WARNING: YOU WILL START TO MISS POWERPOINT

THE PARTY ANIMAL

ADVANTAGES: YOU WILL FINALLY LEARN EFFECT OF BEER ON KIDNEY SECTIONS
WARNING: WILL DRINK YOU UNDER THE TABLE

THE COMEDIAN

ADVANTAGES: OCCASIONALLY FUNNY
WARNING: MAY CRY IF PITY LAUGH NOT GIVEN

THE SEXIST

ADVANTAGES: GREAT IF YOU'RE A FEMALE
WARNING: YOU MAY NOT BE A FEMALE

THE DUMMY

ADVANTAGES: WRITES EASY EXAMS
WARNING: BOARD EXAM WILL BE WRITTEN BY SOMEONE WITH ACTUAL MEDICAL KNOWLEDGE

THE OMNISCIENT

ADVANTAGES: KINDA COOL HOW HE KNOWS SO MUCH
WARNING: SEE FINAL EXAM

THE UNMEMORABLE

ADVANTAGES: NOT MEMORABLY HORRIBLE
WARNING: WILL MAKE UP MOST OF YOUR EDUCATION

THE VISUAL ANALOG SCALE (VAS) FOR PAIN: THE MED STUDENT VERSION

- 😊 0 = 4th YEAR
- 🙂 1 = RADIOLOGY ELECTIVE
- 🙂 2 = STEP 2
- 😐 3 = STEP 2 CLINICAL SKILLS
- 😕 4 = PAYING FOR BOTH STEP 2's
- ☹️ 5 = POST-EXAM HANGOVER
- ☹️ 6 = BREAKING UP WITH ANATOMY LAB PARTNER
- 😢 7 = PSYCH COMORBIDITIES
- 😣 8 = STEP 1
- 😖 9 = RETRACTING POST-CALL
- 😭 10 = (ASSISTING IN) LABOR
- 😭 12 = ASSESSING CHRONIC LOWER BACK PAIN

HOW LONG WILL YOU LIVE?
A FUN TEST FOR MEDICAL STUDENTS*

START THIS TEST WITH 73 POINTS	IF YOU WANT TO BE A SURGEON, SUBTRACT 5	IF YOU HAVE MADE A YOUTUBE VIDEO ABOUT MEDICINE, SUBTRACT 2 ("LIKE A SURGEON")
IF YOU ARE FEMALE, ADD 4	IF YOU WANT TO BE AN OB/GYN, SUBTRACT 7	IF YOU START STUDYING FOR TESTS 12 HOURS BEFORE, SUBTRACT 5
IF MALE, SUBTRACT 5	IF YOU WANT TO DO PSYCH OR PM&R, ADD 5	IF YOU ALREADY HAVE KIDS, SUBTRACT 3 FOR EACH KID ("MOMMYYY!")
IF YOU ARE MARRIED, ADD 2	IF YOU ARE AOA, SUBTRACT 2	IF YOUR PARENTS ARE DOCTORS, SUBTRACT 5 ("NO PRESSURE")
IF MARRIED TO A MED STUDENT, SUBTRACT 2	IF YOU ARE JUNIOR AOA, SUBTRACT 5	IF YOU'RE ALREADY LOOKING AT FELLOWSHIPS, SUBTRACT 3
IF YOU HAVE SAID "I'M NOT HERE TO MAKE FRIENDS!" SUBTRACT 5	IF YOU GO TO SCHOOL IN CALIFORNIA OR HAWAII, ADD 5	IF THIS TEST PISSES YOU OFF, SUBTRACT 8 ("THIS IS UNFAIR")
IF YOU HAVE SAID "SHRUG P=MD" ADD 5	IF YOU STUDY IN LARGE GROUPS, SUBTRACT 2. SUBTRACT 4 IF GROUP IS ALL FEMALE	VIOLA! YOUR SCORE IS YOUR LIFE EXPECTANCY! (HAVE A NICE LIFE)

*RESULTS TO BE PUBLISHED IN JAMA

TIPS FOR USMLE STEP 1.

YOUR WORST SUBJECT WILL COMPRISE 50% OF THE EXAM

THE OTHER 50% WILL BE ON THE FEMALE PELVIS

WHEN IN DOUBT...

THE ANSWER IS PRIAPISM

IF YOU DO BADLY ON THE EXAM, BUILD A TIME MACHINE, GO BACK AND FIX YOUR SCORE

JUST MAKE SURE YOUR MOTHER DOESN'T FALL IN LOVE WITH YOU

BRING EARPLUGS TO PROTECT AGAINST NOISY PEOPLE. BRING NOSEPLUGS TO PROTECT AGAINST SMELLY PEOPLE. ALSO, BRING A BLINDFOLD TO PROTECT AGAINST UGLY PEOPLE "THIS IS PERFECT"

THE SAME #1 BASIC GUIDELINE APPLIES AS TO EVERY OTHER EXAM YOU'VE TAKEN:

NEVER FALL IN LOVE.

AND THERE'S ALWAYS PLAN B:

HIDE UNDER A PILE OF COATS AND HOPE EVERYTHING TURNS OUT OKAY.

ASK A MED STUDENT WHO'S BEEN STUDYING FOR USMLE STEP 1:

"MED STUDENT WHO'S BEEN STUDYING FOR STEP 1, DO YOU THINK 2 YEARS IS LONG ENOUGH TO DATE BEFORE MARRYING?"

"2??? YOU MEAN LIKE NEUROFIBROMATOSIS 2, WHICH IS LOCATED ON CHROMOSOME 22?"

"OR LIKE MECKEL'S DIVERTICULUM, WHICH IS LOCATED 2 FEET AWAY FROM THE ILEOCECAL JUNCTION AND OCCURS IN 2% OF THE POPULATION?"

"UH..."

"MAYBE THE ANSWER IS IN STEP 1 FIRST AID. LET'S CHECK."

MED SCHOOL CLINICAL YEARS: BEFORE + AFTER

BEFORE

AFTER

BEFORE

AFTER

BEFORE

AFTER

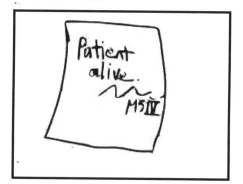

STAGES OF THE MED. SCHOOL CLINICAL YEARS

1: CLUELESS

2: ENTHUSIASTIC

3: DEPRESSED

4: SEASONED

5: CHECKING OUT

6: CHECKED OUT

CRUEL RESIDENT STORIES

REASONS WHY IT'S BETTER TO BE A DOG THAN A MED STUDENT:

MORE FRESH AIR:

 vs.

LOTS OF TREATS

 vs.

SIMPLE INSTRUCTIONS:

 vs.

ALLOWED TO BITE BAD PEOPLE ON ASS:

 vs

WHAT TO SAY WHEN YOUR ATTENDING ASKS WHY YOU WERE LATE:

FIVE WAYS FOR A MEDICAL STUDENT TO CUT A SURGICAL KNOT:

TOO SHORT — TOO LONG — TOO TWIRLY — DREDS — PUBIC

THINGS MED STUDENTS ARE GOOD AT:

RECTAL EXAMS (OF COURSE)

CUTTING KNOTS TOO LONG/SHORT
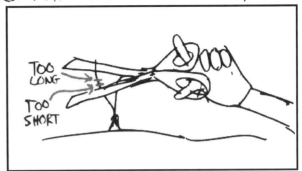

REPOSITORY FOR 4 BY 4'S / PENS

PICKING UP YOUR DRYCLEANING

IDENTIFY THE MED STUDENT ROTATION BASED ON THEIR HANDS:

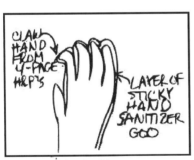
MEDICINE

SURGERY

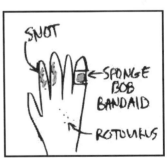
PEDIATRICS

PSYCHIATRY

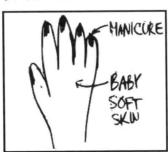
DERM ELECTIVE

OB/GYN

TIPS FOR 3rd YEAR OF MED SCHOOL

THE CORRECT ANSWER TO "WOULD YOU LIKE TO DO IT?" IS:

IT DOESN'T MATTER WHAT "IT" IS. ("IT" IS PROBABLY A RECTAL EXAM.)

THE PROPER GREETING FOR FELLOW MS3'S:

IF YOU NEED TO GO TO A DOCTOR, YOU'LL PROBABLY SEE ONE OF YOUR ATTENDINGS

DON'T HAVE A UROLOGIC, GI, OR GYNECOLOGIC PROBLEM.

YOU MIGHT WORRY ABOUT KILLING SOMEONE...

DON'T WORRY. YOU'LL NEVER EVEN CLOSE TO HAVE AN OPPORTUNITY.

THINGS YOU CAN DO AS A MED STUDENT NOT TO PISS OFF YOUR OB/GYN RESIDENT:

THINGS THAT ARE GUARANTEES FOR YOUR THIRD YEAR OF MED SCHOOL:

AT SOME POINT, YOU WILL GET DRENCHED IN A BODY FLUID

YOUR PENLIGHT BATTERY WILL DIE, DESPITE HAVING ONLY USED IT, LIKE, ONCE

ALL YOUR OB PATIENTS WILL BE YOUNGER THAN YOU AND HAVE MANY MORE KIDS

DURING YOUR PEDS ROTATION, YOU WILL GET A GI INFECTION

CRUEL RESIDENT STORIES: SEEING PRACTICE PATIENTS AS PART OF MY INTRO TO CLINICAL MED COURSE AS A 2ND YEAR MED STUDENT....

NICKNAMES YOU WANT TO AVOID DURING YOUR MED SCHOOL CLERKSHIPS

- THE COFFEE BITCH
- RECTAL EXAM GIRL
- THE DUMB ONE
- THE FAT ONE
- SPANKY
- THE ONE WHO LOST CONTROL OF HER BOWELS DURING THAT C-SECTION
- HEP C GIRL
- WE HAVE A MED STUDENT?

MEDICAL STUDENT Q&A:

YOU HAVE JUST BEEN KICKED OUT OF A SURGERY AFTER 2 MINUTES FOR BREAKING STERILE. ONCE OUTSIDE, YOU SEE THE SPONGE YOU JUST SCRUBBED WITH HOVERING ON TOP OF THE TRASH BIN, UNTOUCHED. DO YOU:

a) WASTEFULLY OPEN NEW SPONGE AND SCRUB AGAIN
b) TAKE SPONGE OFF GARBAGE AND SCRUB WITH IT AGAIN
c) QUIT MEDICINE AND GET JOB FOLDING JEANS AT GAP

STAGES OF NEEDING TO PEE DURING AN 8-HOUR SURGERY:

HOUR 1: FINE

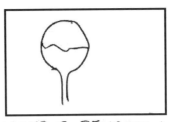
HOUR 3: SENSATION OF BLADDER FULLNESS → CONTINENCE ACHIEVED THROUGH EXTERNAL URETHRAL SPHINCTER

HOUR 4: LEG CROSSING NOTED

HOUR 5: DAYDREAMING ABOUT TOILETS

HOUR 7: THE TOILET MIRAGE

HOUR 8: RELIEF VS. "OH SHIT!"

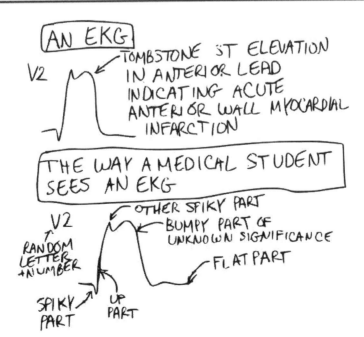

WAYS MED SCHOOL IS LIKE KINDERGARTEN

YOU NEED TO ASK PERMISSION TO PEE:

LEARNING TO USE SCISSORS:

FREQUENT BUT CONSTRUCTIVE SCOLDINGS:

NAPS:

MEDICAL STUDENTS VS. LAW STUDENTS

MED STUDENTS

PHONY IDEALISM:

WILLING TO DO DIRTY WORK:

THINKS HE'S GOD:

GREAT FIRST JOB:

LAW STUDENTS

PHONY IDEALISM:

CAN'T WAIT TO DO DIRTY WORK:

THINKS HE'S EARL WARREN:

FLEXIBLE FIRST JOB:

Scoring:
<10: You are not a nerd. Except you sort of still are, since you're in med school.
11-25: You are a cool nerd. Like Harold and Kumar in Harold and Kumar Go to White Castle. Even though you're kind of nerdy, you can still get girls, get high, and make an epic journey to get those delicious tiny burgers.
26-40: You are an intellectual nerd. Other intellectual nerds include Al Gore or David Souter. You're super nerdy, but people still respect you enough to elect you Vice President of Neurology.
>40: You are an uber-nerd. Other uber-nerds include Mark Zuckerberg or those guys from Revenge of the Nerds. Although everyone always kind of made fun of you, not so much after you become the world's youngest billionaire.

WHAT IS MATCH DAY?

MED SCHOOL CLASSES ARE SUSPENDED FOR A TENNIS COMPETITION

EVERYONE DRESSES IN MATCHING PANTS SUITS

YOU GET TO TRY TO SET FIRE TO YOUR MED SCHOOL

WHEN YOU LOOK INTO THE FLAME OF A MATCH, YOU SEE THE HORRIBLE IMAGE OF YOUR FUTURE CAREER

The Science Behind the Residency Match

THE 13 TYPES OF ANATOMY LAB GROUPS

THE SMELLY BODY GROUP

ADVANTAGE: YOU WILL NEVER BECOME THE CROWDED GROUP
DISADVANTAGE: PEOPLE MAY SUSPECT IT'S NOT THE CADAVER CAUSING THAT SMELL.

GROUP WHERE ONLY ONE PERSON SHOWS UP

ADVANTAGE: YOU LEARN A LOT
DISADVANTAGE: MOST OF WHAT YOU LEARN IS TO PICK BETTER PARTNERS

THE ALL-FEMALE GROUP

ADVANTAGE: FUN TO LOOK AT
DISADVANTAGE: MAY NOT RESEMBLE PICTURE

THE ALL-MALE GROUP

ADVANTAGE: GET A LOT OF WORK DONE
DISADVANTAGE: NOT WORTH IT

GROUP THAT IS STILL AFRAID OF BODY

ADVANTAGE: YOU'VE GOT MORE SENSITIVITY THAN ANY OF YOUR CLASSMATES
DISADVANTAGE: YOUR CLASSMATES WILL THINK YOU'RE A LITTLE WUSS

GROUP WHERE ONE MEMBER DRESSES AS COW

ADVANTAGE: REMAINS FUNNY LONGER THAN EXPECTED
DISADVANTAGE: UDDERS DO NOT ACTUALLY GIVE MILK

UNPREPARED GROUP

ADVANTAGE: REQUIRES LITTLE PREPARATION TIME
DISADVANTAGE: MAY TURN INTO SLOW GROUP

THE PERFECT GROUP

ADVANTAGE: MAY GET TO COME BACK NEXT YEAR TO T.A.
DISADVANTAGE: MAY HAVE TO COME BACK NEXT YEAR TO T.A.

GROUP THAT NEVER FIGURED OUT COMBINATION TO ANATOMY LAB

ADVANTAGE: ORIGINAL PAIR OF GLOVES LASTS THROUGH ENTIRE SEMESTER
DISADVANTAGE: THE OTHER STUDENTS ARE LAUGHING AT YOU.

MUTANT BODY

ADVANTAGE: IF YOU'RE LUCKY, YOUR CADAVER MAY BE WRITTEN UP IN THE NEW ENGLAND JOURNAL
DISADVANTAGE: YOU'RE NOT LUCKY

INCOMPETENT GROUP

"OOPS! I JUST CUT THE FACIAL ARTERY!"

"NAW, IT'S PROBABLY JUST FASCIA."

ADVANTAGE: LAB GOES SURPRISINGLY QUICK.
DISADVANTAGE: MAY BE INDICATIVE OF THE REST OF YOUR LIFE

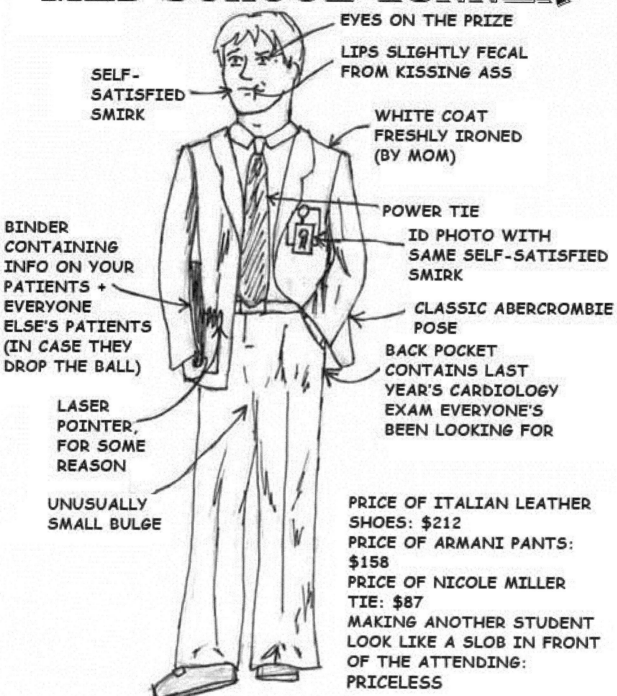

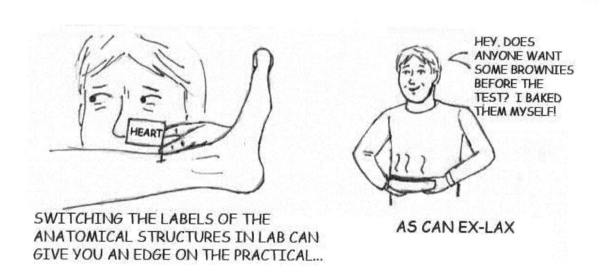
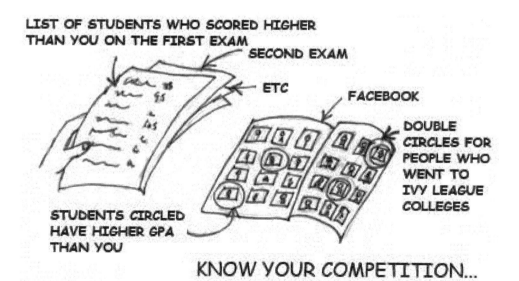

KEEP YOUR CLASSMATES INFORMED OF YOUR SUCCESSES

BUT IF SOMEONE DOES BETTER THAN YOU, DON'T PANIC

IT WILL ALL RESOLVE ITSELF SOMEHOW

3
RESIDENCY

RESIDENCY INTERVIEWS: A WINDOW INTO THE INTERVIEW ROOM

NEUROLOGY

ORTHOPEDIC SURGERY

PSYCHIATRY

PM&R

PRIMARY CARE

DERMATOLOGY

GUARANTEES FOR YOUR FIRST WEEK OF INTERNSHIP

YOU WILL GET TRAPPED IN A STAIRCASE*

YOU WILL BE REFERRED TO IN THIRD PERSON

YOU WILL RECEIVE A LOOK OF ASTONISHMENT AT YOUR UTTER STUPIDITY

YOU WILL BE REDUCED TO TEARS TRYING TO FILL OUT A FORM

YOU WILL THINK TO YOURSELF: "I DON'T WANT TO BE A DOCTOR!"

YOU WILL GO ON DESPERATE HUNT FOR GUAIAC CARD/FLUID

YOU WILL GET CALLED ON EVERY. ORDER. YOU. WRITE.

IT WILL SEEM RIDICULOUS THAT EVERYONE IS CALLING YOU DOCTOR

EVERY 6 SECONDS, YOU WILL FANTASIZE ABOUT QUITTING

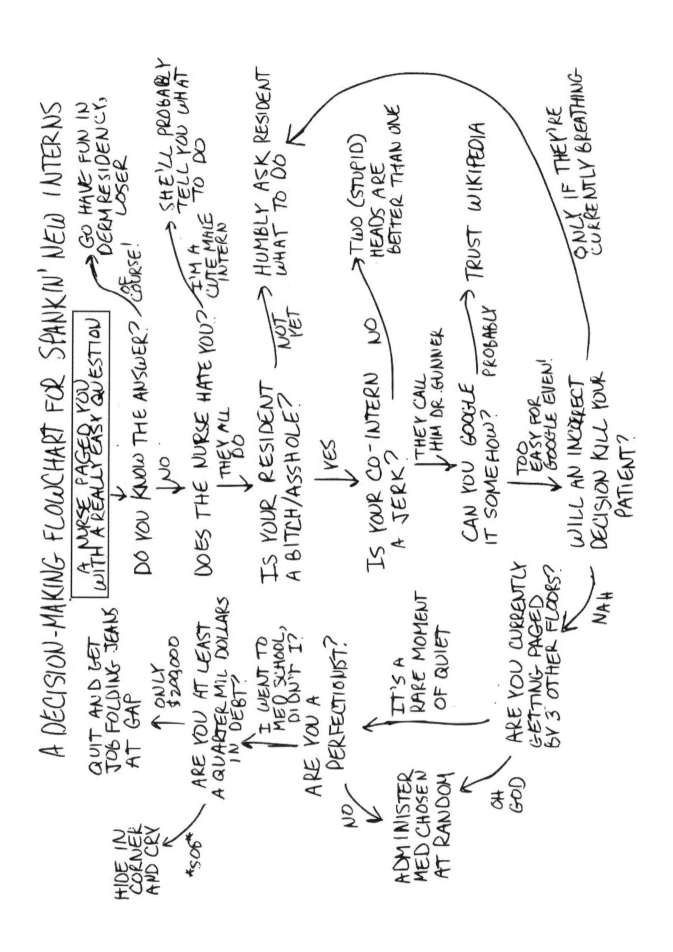

THINGS I LEARNED DURING INTERN YEAR

THE FIRST 2 TIMES YOUR PAGER GOES OFF ARE EXCITING

EVERY TIME AFTER THAT IS PROGRESSIVELY MORE GUT-WRENCHING

WHOEVER YOUR CO-INTERN IS WILL BE YOUR NEW BFF...

OR YOUR MOST HATED ENEMY

THERE IS AN INVERSE RELATIONSHIP BETWEEN AMOUNT OF CRAP IN YOUR POCKET AND LEVEL OF TRAINING

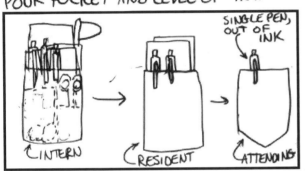

WHEN YOU GIVE A PRESENTATION, EVERYONE IN THE ROOM WILL KNOW MORE ABOUT YOUR TOPIC THAN YOU DO

YOU WILL NEVER ESCAPE THE HOSPITAL, EVEN IN SLEEP

FREE FOOD IS THE GREATEST JOY IN LIFE

WHITE COATS STAY WHITE ~5 MINS

THERE IS NO PRIVACY, EVEN IN THE BATHROOM

WHEN IN DOUBT, TYLENOL OR STOOL SOFTENERS WILL USUALLY BUY YOU SOME BREATHING ROOM

REVELATIONS FROM INTERN YEAR

SLEEP IS NOT NEEDED TO LIVE....
NOR IS IT NEEDED TO MAKE LIFE OR DEATH DECISIONS ABOUT PATIENTS

PATIENTS ARE PRETTY HARD TO KILL

DALE DUBIN IS AN (ALLEGED) CHILD PORNOGRAPHER

THE MONTHS OF INTERNSHIP

JULY: DANGEROUS

AUGUST: DEPRESSED

SEPTEMBER: OVERWHELMED

OCTOBER: DEPRESSED

NOVEMBER: HOPEFUL

DECEMBER: DEPRESSED

JANUARY: TIRED

FEBRUARY: DEPRESSED

MARCH: JADED

APRIL: DEPRESSED

MAY: FRIGHTENED

JUNE: DEPRESSED

TYPES OF PAGES FROM NURSES DURING THE NIGHT

THE "3AM CONSTIPATION ATTACK" PAGE

THE "YOUR PATIENT CAN'T SLEEP SO NEITHER CAN YOU" PAGE

THE "VAGUE PATIENT REQUEST" PAGE

THE "NO INFO" PAGE

THE "SHOELESS JOE" PAGE

THE "STUPIDEST PAGE EVER" PAGE

THE "FLIPPING THROUGH THE CHART AT 3AM" PAGE

THE "FROOT LOOPS?! ARE YOU FUCKING SERIOUS??" PAGE

THE ACTUAL, REAL, SCARY PAGE

CRUEL RESIDENT STORIES
BACKGROUND: MY FIRST OVERNIGHT CALL AS AN INTERN

HOUR 1

HOUR 10

HOUR 20

HOUR 32

CRUEL RESIDENT STORIES

*IN CASE YOU CAN'T DO THE MATH, THIS MEANS I WOULD START PREROUNDING AT 3AM EVERY DAY

WISDOM FROM MY FIRST YEAR OF PM&R RESIDENCY:

IN BACK/KNEE/SHOULDER PAIN, YOU HAVE SEVERAL OPTIONS:

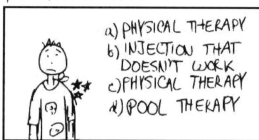

BEING 8 MONTHS PREGNANT IS NOT AN EXCUSE NOT TO DO ANYTHING...

INCLUDING HOLDING A PATIENT'S HEAD DURING X-RAYS

PATIENTS WILL SOMETIMES BREAK THEIR WEIGHTBEARING PRECAUTIONS TO CHASE YOU

DEFINITION OF DIFFICULT FAMILY MEMBER:

WHATEVER FAMILY MEMBER YOU'RE TALKING TO

YOU CAN COVER FOR SOMEONE A DOZEN TIMES...

BUT THE ONE TIME THEY COVERED FOR YOU WILL BE ALL THEY REMEMBER.

EVERY BOWEL MEDICATION KNOWN TO MAN

5 Phrases I Overused as a Resident

Post-Calling

Definition: Making calls to exes due to exhaustion-lowered inhibitions after being awake 32 hours in a row.

THE 10 TYPES OF RESIDENTS

DR. KISS ASS

DR. LAZY BUM

DR. SCHEMER

DR. PREGGO

DR. CHIEF

*Actually said by my chief resident

DR. OVERENTHUSIASTIC

DR. EFFICIENT

DR. CASANOVA

DR. WORRIED

DR. MISERABLE

EMBARRASSING RESIDENCY STORIES: STUCK IN A STAIRWELL!

9 MONTHS INTO MY INTERN YEAR, I TOOK AN ALTERNATE STAIRCASE UP ONE FLIGHT TO SEE A PATIENT

I GOT UP TO THE 5th FLOOR AND DISCOVERED THE STAIRS WERE LOCKED

I WENT BACK TO THE 4th FLOOR AND THAT DOOR WAS LOCKED TOO!

LUCKILY, I HAD MY CELL PHONE AND CALLED MY RESIDENT

BUT AFTER 10 MINUTES, MY RESIDENT COULDN'T FIND ME

THAT'S WHEN I REALIZED I HAD ACCIDENTALLY GONE UP INSTEAD OF DOWN

I WENT BACK DOWN TO THE 4th FLOOR AND THE DOOR WAS UNLOCKED ALL ALONG

WHAT TO DO IF YOU GET SICK DURING RESIDENCY

STEP 1: GO TO WORK

STEP 2: ILLUSTRATE ILLNESS TO ATTENDING

STEP 3: AWAIT RESPONSE

STEP 4: DISSEMINATE ILLNESS

STEP 5: PROVE SELF WORTHLESS

STEP 6: CALL CHIEF TO ARRANGE OWN COVERAGE

STEP 7: KEEP WORKING UNTIL COVERAGE ARRIVES

INTERVIEW TIPS:
RESIDENCY PROGRAM RED FLAGS:

CRYING RESIDENTS

>50% SLEEPING RATIO DURING MORNING REPORT

SIGNS OF CRUEL OR UNUSUAL PUNISHMENT

ACCEPTABLE REASONS FOR AN INTERN TO CALL IN SICK

PROJECTILE VOMITING

APPENDIX FLYING OUT OF ABDOMEN AND EXPLODING MIDAIR

SEPTIC SHOCK REQUIRING INTUBATION

DEATH

CRUEL RESIDENT STORIES

EMBARRASSING RESIDENCY STORIES

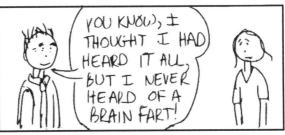

WAYS TO HELP REMEMBER THINGS ON WARDS

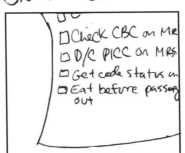

CHECKBOXES GALORE

WRITING ON HANDS

MAKE UP A RAP

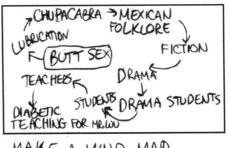

MAKE A MIND MAP

How to tell your attending isn't listening to your patient presentation:

Sings Journey under his breath

Interrupts repeatedly to tell fishing stories

Answers all questions the same way

Staring at nursing student's breasts

Signs your med student has boundary problems:

Copies everything you do

Watches you write your notes

Follows you into bathroom*

Tries to friend you on Facebook

*Bonus pts if he tries to follow you into stall

WAYS BEING ON CALL IS LIKE BEING AT A SLEEPOVER:

EVERYONE WEARS PJs:

YOU EAT TONS OF JUNK FOOD:

AT SOME POINT, YOU CALL SOMEONE A BITCH:

SOMEONE TELLS YOU WHEN YOU CAN + CAN'T SLEEP

CRUEL RESIDENT STORIES
SETTING: 3AM ON A CALL NIGHT DURING INTERNSHIP

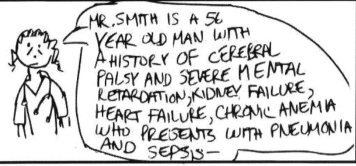

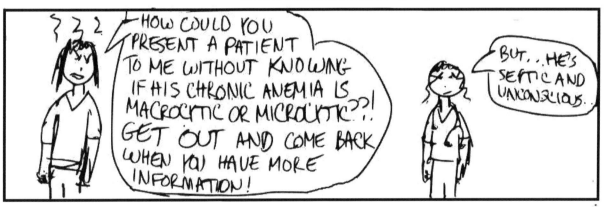

STAGES OF A 30 HOUR OVERNIGHT CALL

HOUR 0: BRIGHT EYED

HOUR 1: DEPRESSED

HOUR 5: BUSY

HOUR 10: BUSIER

HOUR 13: FULL

HOUR 14: REMORSEFUL

HOUR 20: SLEEPY

HOUR 23: DRUNK

HOUR 26: OUTER BODY

HOUR 30: JUBILANT

HOUR 31: SLEEPING

RESIDENT SPEAK: TYPES OF WEEKENDS

GOLDEN WEEKEND: HAVING ENTIRE WEEKEND FREE

SILVER WEEKEND: HAVING ONE DAY OF WEEKEND OFF AND NO CALL

BEIGE WEEKEND: FRIDAY OR SUNDAY OVERNIGHT CALL

BROWN WEEKEND: SATURDAY NIGHT CALL, OR FRIDAY + SUNDAY OVERNIGHT CALL WHOLE GODDAMN WEEKEND RUINED

CROSS COVER QUESTIONS:
"DOCTOR, MR. SMITH ONLY PEED 20cc IN THE LAST TWO HOURS. WHAT WOULD YOU LIKE TO DO ABOUT THAT?"

SMARTASS ANSWER #1: "YEAH? WELL I PEED ZERO CC'S IN THE LAST 8 HOURS!"

SMARTASS ANSWER #2: "MAYBE HE NEEDS SOME PRIVACY? LET'S SEND HIM HOME!"

SMARTASS ANSWER #3: "ISN'T BEER A DIURETIC?"

ACTUAL ANSWER:

"UHHHH..." — SEARCHES DESPERATELY FOR SIGN OUT CARD, WHICH IS BLANK (OF COURSE)

BORED AT WORK?
TRY THE DICTATING RESIDENT DRINKING GAME!
HERE'S HOW TO PLAY!
TAKE ONE SHOT EACH TIME THE RESIDENT SAYS...

TAKE A DOUBLE SHOT EACH TIME THE RESIDENT...

PAUSES DICTATION AND SEARCHES FRANTICALLY THROUGH CHART

MUMBLES NAME OF MED, HOPING TRANSCRIPTIONIST WILL FIGURE OUT WHAT IT IS

GETS PAGED

TAKE A TRIPLE SHOT EACH TIME THE RESIDENT...

FORGETS CRUCIAL PIECE OF INFORMATION

APOLOGIZES PROFUSELY TO TRANSCRIPTIONIST

BY NOW YOU SHOULD BE GOOD AND DRUNK...
ENJOY WORK!

EMBARRASSING RESIDENCY STORIES

AS A RESIDENT, I USED TO ASSIST IN THE OR DOING INJECTIONS UNDER FLUORO

I LOOKED DOWN AND....

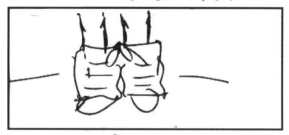

MY SCRUB PANTS WERE AROUND MY ANKLES!!

BUT I MANAGED TO GET MY PANTS UP WITHOUT THE ATTENDING NOTICING. I ALMOST GOT AWAY WITH IT TOO, IF NOT FOR THE STUPID PAIN FELLOW...

ONE DAY WHEN I WAS FILLING THE SYRINGES, I NOTICED MY LEGS COULDN'T MOVE THAT WELL

FORTUNATELY, MY SECRET WAS CONCEALED BY MY LEAD SKIRT. BUT THE ATTENDING FREAKED.

CRUEL RESIDENT STORIES: THE REVENGE OF THE INTERN

LATER, IN PATIENT'S ROOM:

FELLOWSHIP

"I'm going to graduate from residency soon. I'm just wondering: should I do a fellowship?"

SHOULD YOU DO FELLOWSHIP?
A CHECKLIST:
DO YOU...
- ☐ ENJOY ACCUMULATING DEBT
- ☐ STILL FEEL TERRIFIED OF MAKING DECISIONS ON YOUR OWN?
- ☐ WANT TO BE THE WORLD'S LEADING EXPERT ON SOMETHING REALLY OBSCURE
- ☐ THINK YOUR CV IS A LITTLE FLIMSY
- ☐ FEEL YOU LEARNED NOTHING IN ALL OF MED SCHOOL AND RESIDENCY
- ☐ HATE YOUR PRIMARY SPECIALTY AND HOPE YOU MIGHT HAVE BETTER LUCK WITH SOMETHING TANGENTIALLY RELATED.

IF YOU CHECKED ONE OR MORE OF THE ABOVE BOXES, WELCOME TO THE WONDERFUL WORLD OF FELLOWSHIP TRAINING! IT NEVER ENDS, DOES IT?

4

ATTENDINGS

7 TYPES OF ATTENDINGS

DR. ATHLETE

ADVANTAGES: WILL GET YOU IN SHAPE
WARNING: MAY ORDER STRESS TEST ON YOU

DR. MOM

ADVANTAGES: ENDLESS SUPPLY OF SPONGEBOB BANDAIDS
WARNING: MAY PUT YOU IN BUBBLE BATH

DR. LIVES AT HOSPITAL

ADVANTAGES: YOU'RE NEVER ALONE
WARNING: NEVER LEAVES YOU ALONE!

DR. KNOW-IT-ALL
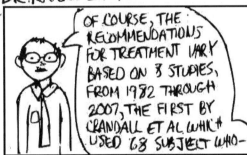
ADVANTAGES: KNOWS EVERYTHING
WARNING: KNOWS YOU KNOW NOTHING

DR. CLUELESS

ADVANTAGES: MAKES YOU FEEL SMART
WARNING: MAY KILL YOUR PATIENTS

DR. ASSHOLE

ADVANTAGES: TEACHES YOU TO HOLD IN YOUR TEARS
WARNING: NEVER RETIRES

DR. NICE GUY

ADVANTAGES: MAKES LIFE WORTH LIVING AGAIN
WARNING: DOESN'T ACTUALLY EXIST

STYLES OF ATTENDING ROUNDS

DR. SPEEDY GONZALEZ

DR. COMPASSIONATE

DR. DETAILS

DR. ANAL

DR. PIMPSTER

DR. EVERYONE NEEDS TO POOP TODAY

INSANE ATTENDING STORIES:
DURING MY PEDIATRIC ENDOCRINOLOGY ROTATION,
MY ATTENDING HAD A THICK ACCENT

IMPORTANT LESSON FROM MED SCHOOL #29:
NEVER TOUCH THE BRAIN

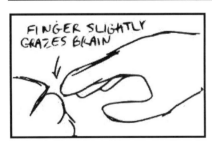

LESSON: NEVER TOUCH THE BRAIN.
EVEN IF THEY TELL YOU IT'S OK, JUST DON'T
DO IT, MAN. IT'S NOT WORTH IT.

INSANE ATTENDING STORIES: PELVIC ABNORMALITIES

*TWO UTERUSES

INSANE ATTENDING STORIES

AN ATTENDING'S GUIDE TO CLINICAL RESEARCH

STEP 1: COME UP WITH CONNECTION SO OBVIOUS NOBODY HAS EVER LOOKED INTO IT BEFORE

STEP 2: FIND RESIDENT OR MED STUDENT TO ~~WRITE IRB, GET FUNDING, RESEAR~~ DO EVERYTHING

STEP 3: THINK OF JOURNAL LEAST LIKELY TO REJECT ARTICLE

STEP 4: ADDRESS LONG LIST OF SCATHING REVIEWER CRITIQUES

STEP 5: CONTEMPLATE FIRST AUTHORSHIP

STEP 6: BASK IN GLORY OF YOUR PUBLICATION

MUSIC STYLINGS OF THE OR ATTENDING

DR. DEATHLY SILENCE

DR. DEATH METAL

DR. JUSTIN BIEBER

DR. ONLY OWNS ONE CD

INSANE ATTENDING STORIES

BACKGROUND: AS A RESIDENT, I WORKED WITH AN ATTENDING WHO REALLY WANTED US TO CALL MEDICATIONS BY THEIR GENERIC NAMES

INSANE ATTENDING STORIES

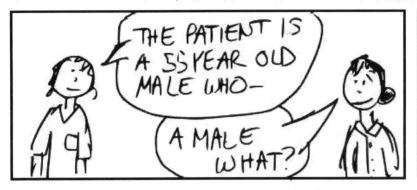

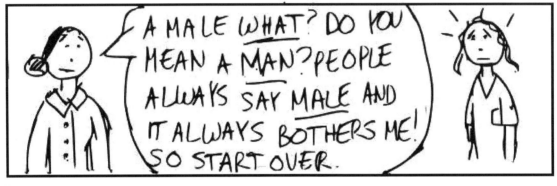
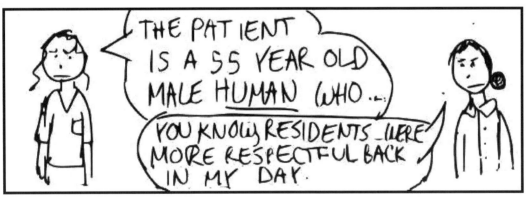

INSANE ATTENDING STORIES: IN THE ICU...

LATER, IN MR. GOMEZ'S ROOM:

5

HAIKUS

AEROSMITH: A HAIKU

LOVE DURING MED SCHOOL
LIVIN' IT UP WHILE YOU STINK
OF FORMALDEHYDE

HAIKU

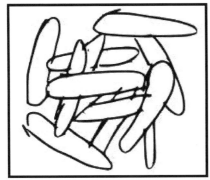

TUBERCULOSIS
HAS SO MANY SYLLABLES
HARD TO WRITE HAIKU

SECOND YEAR MED SCHOOL (A HAIKU)

SECOND YEAR IS NOT
AS EASY AS THEY SAID IT
WOULD BE. PLEASE KILL ME.

A HAIKU

CODE BLUE ON PATIENT

INTERN RUNS TO THE BEDSIDE

CODE BLUE ON INTERN

MY FIRST PHARMACOLOGY EXAM: A HAIKU

HOW CAN IT BE THAT
ACETAMINOPHEN IS
ONE THIRD OF EXAM?

MED STUDENT IN THE ER: A HAIKU

MAN IN STRETCHER CODES
THE DOCTOR YELLS, "INTUBATE!"
I CHEW MY TWIX BAR

6

THE SPECIALTIES

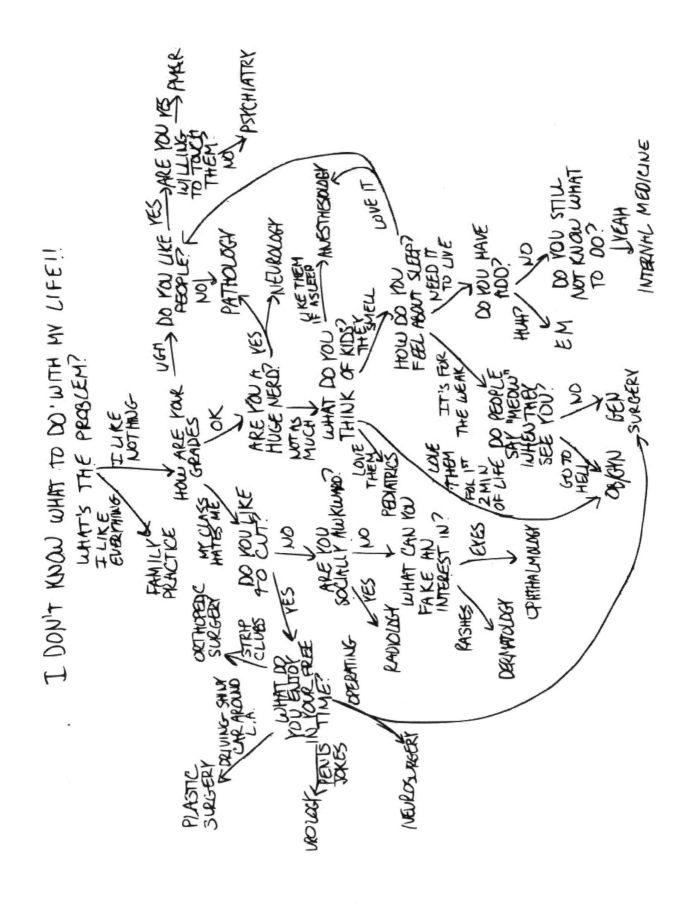

HIGH SCHOOL STEREOTYPES...
OF MEDICAL SPECIALTIES

PATHOLOGIST: THE GOTH
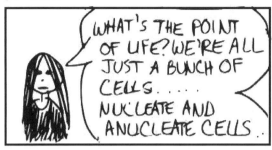

NEUROLOGIST: THE NERD

UROLOGIST: THE CLASS CLOWN

ORTHOPOD: THE JOCK

OB/GYN: THE CHEERLEADER

MALE OB/GYN: GUY WHO WATCHES CHEERLEADERS IN SHOWERS

DERMATOLOGIST: TEACHER'S PET

DENTIST: THE PARTY ANIMAL

EVOLUTION OF SPECIALTY CHOICE

1st YEAR

2nd YEAR

3rd YEAR: MEDICINE CLERKSHIP

3rd YEAR: PEDS CLERKSHIP

4th YEAR

RESIDENCY

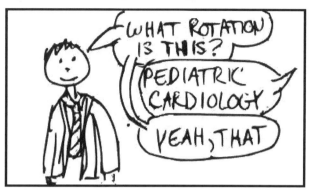

7

EVERYTHING ELSE

FUN WITH MEDICAL ABBREVIATIONS

NAD: GENITALIA (SINGULAR)

NPH: NEIL PATRICK HARRIS

F/U: FUCK UP

ICP: INSANE CLOWN POSSE

PERRLA: A PRETTY NAME FOR A GIRL

SOB: DUH

* NAD: NO APPARENT DISTRESS
  NPH: NORMAL PRESSURE HYDROCEPHALUS
  PERRLA: PUPILS EQUAL, ROUND, REACTIVE TO LIGHT + ACCOMMODATION
  SOB: SHORTNESS OF BREATH
  F/U: FOLLOW-UP
  ICP: INTRACRANIAL PRESSURE

APRIL FOOLS JOKES THAT ARE NOT FUNNY

THE CROSS COVER FAKE OUT:

THE EXAM FAKE OUT:

THE L&D FAKE OUT:

A few more:

What the doctor says: "Can you point with one finger to where your pain is?"
What the patient hears: "Can you wave your hand around in a huge circle to tell me where your pain is?"

What the doctor says: "What is your pain level on a scale of 1 to 10?"
What the patient hears: "Can you review your pain level with every possible activity you can think of on a scale of 1 to 20."

What the doctor says: "What medications are you taking?"
What the patient hears: "Can you describe the color and size of no more than half of the pills you're taking?"

What the doctor says: "Do you have any other medical problems?"
What the patient hears: "Can you give me a detailed history of how all your other medical problems were diagnosed? And if you have any major medical problems, like diabetes, I don't need to know about those."

THINGS THAT REALLY HAPPEN DURING A BOOB JOB:

ALL THE INSTRUMENTS HAVE "BREAST" IN THEIR NAMES

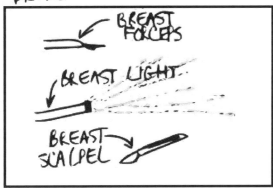

THERE IS A "TEST BOOB" FILLED WITH SALINE

THE SLEEPING PATIENT IS SAT UP AND EVERYONE GATHERS 'ROUND TO LOOK

THE MED STUDENT MAKES A REALLY DUMB COMMENT

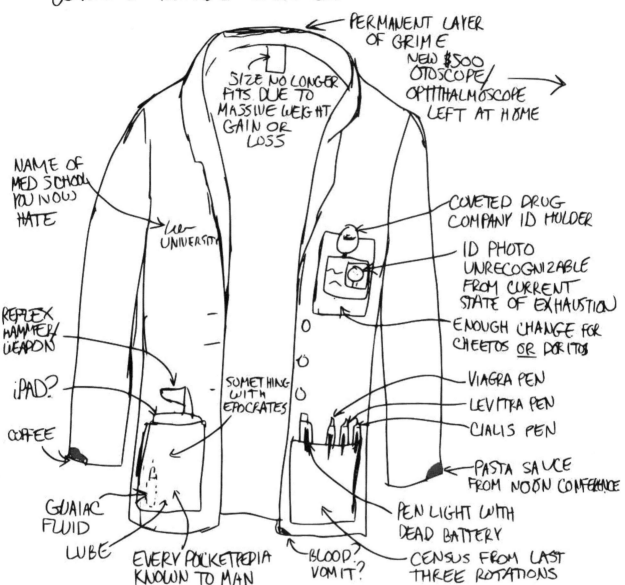

SIGNS YOU MIGHT HAVE HAD TOO MUCH CAFFEINE

ESSENTIAL TREMOR

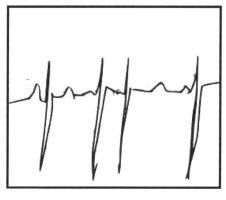

CARDIAC ARRHYTHMIA

FEET NO LONGER TOUCHING GROUND

LAMEST RESIDENT HALLOWEEN COSTUMES

A DOCTOR

ANYTHING ANATOMICAL

SEXY CAT

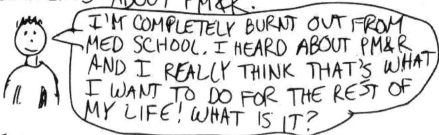

AS A YOUNG PHYSIATRIST, I OFTEN GET QUESTIONS FROM MED STUDENTS ABOUT PM&R:

"I'M COMPLETELY BURNT OUT FROM MED SCHOOL. I HEARD ABOUT PM&R AND I REALLY THINK THAT'S WHAT I WANT TO DO FOR THE REST OF MY LIFE! WHAT IS IT?"

(WE LOVE THIS QUESTION)

SO...
WHAT DOES A PHYSIATRIST DO?
WE SEE A WIDE VARIETY OF PATIENTS:

SPINAL CORD INJURY   TRAUMATIC BRAIN INJURY   MUSCULOSKELETAL PAIN

WE ALSO DO A WIDE VARIETY OF PROCEDURES:

BOTOX FOR SPASTICITY   EPIDURAL STEROID INJECTIONS UNDER FLUORO   ELECTROMYOGRAPHY

WELL, I THINK YOU CAN SEE THAT PM&R IS A PRETTY AWESOME FIELD. IN ANY CASE, IF YOU HAVE ANY QUESTIONS, YOU KNOW WHERE TO FIND US:

(JUST KIDDING, WE WORK PRETTY HARD.)
(AT GOLF.)

TOP REASONS TO CALL A PM&R CONSULT

PATIENT IS REALLY FAT

PATIENT BEING DISCHARGED SOON AFTER LONG HOSPITAL STAY, HOPING PM&R H&P CAN SERVE AS D/C SUMMARY

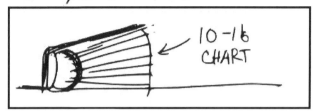

FRIDAY AFTERNOON, NEED TO DIURESE SERVICE

PAIN, SOMEWHERE

WAS AUTO-CHECKBOX ON STROKE PROTOCOL ORDERS

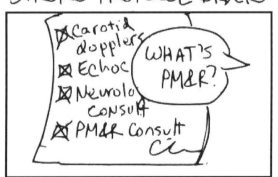

THOUGHT IT WAS PHYSICAL THERAPY

4 YEARS OF MED SCHOOL TUITION:

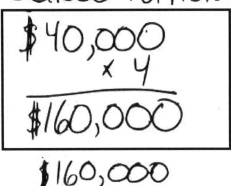

$160,000

4 YEARS OF RENT FOR A SINGLE UNFURNISHED BEDROOM:

$40,000

STEP 1 + STEP 2 CK + STEP 2 CS + STEP 3 + QBANK:

$3,000

1,000 PACKAGES OF RAMEN NOODLES

$1,000

HAVING A JOB THAT PAYS LESS THAN MINIMUM WAGE WHEN YOU'RE $200,000+ IN DEBT:

PRICELESS

THERE ARE SOME THINGS MONEY CAN'T BUY. FOR EVERYTHING ELSE, THERE'S CRUSHING STUDENT DEBT

ORBICULARIS ORIS

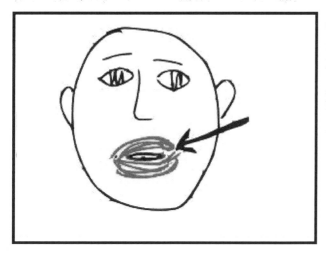

DEF'N: MUSCLE AROUND MOUTH RESPONSIBLE FOR CLOSING AND PUCKERING LIPS

ORBICULARISAURUS

DEF'N: DINOSAUR THAT ROAMED EARTH DURING JURASSIC ERA.

STAGES OF WANTING TO BE A DOCTOR:

AGE 3: PRESCHOOL

AGE 6: GRADESCHOOL

AGE 10: MIDDLE SCHOOL

AGE 15: HIGH SCHOOL

AGE 20: MCATs

AGE 22: INTERVIEWS

USE #149 FOR SURGICAL GLOVES:

TAKING YOUR HOUSE ON A MAGICAL ADVENTURE AS INSPIRED BY "UP"

DESCRIBE YOUR HEADACHE:

WHAT IS THE QUALITY OF YOUR HEADACHE?

- PULSATING
- SHARP
- ELECTRIC
- TICKLISH — HEH

WHAT IS THE DURATION OF YOUR HEADACHE?

- 60 SECONDS
- FOR FREAKING EVER
- UNTIL HUSBAND STOPS WANTING SEX

WHERE IS YOUR HEADACHE LOCATED?

- UNILATERAL
- BILATERAL
- IN HAIR
- GROIN*

*REPRESENTS 1% OF HEADACHES

DO YOU HAVE ANY ASSOCIATED SYMPTOMS?

- VOMITING
- PHOTOPHOBIA
- SHOOTING WEBS (GO WEB!)
- TIME TRAVEL

DO YOU SEE ANY VISUAL AURAS?

- ZIGZAGS
- DEAD PEOPLE (← BRUCE WILLIS)
- LARGE SCARY BUNNY

THE JOYS OF HOME CALL: STAGES OF GETTING PAGED AT 2AM

1: PEACEFUL

2: ADRENALINE RUSH

3: OPTIMISM

4: SINKING FEELING

5: CURSING/DRESSING

6: POTENTIAL MVA

7: PARKING ACROSS 3 HANDICAPPED SPACES

8: HERO MOMENT

9: HORROR

10: BLIND FURY

11: SPEEDING THROUGH SOUND BARRIER

12: TOO PISSED OFF TO SLEEP

THREE JOKES FOR KIDS:
MY OUTPATIENT PEDIATRICS ATTENDING USED TO ALWAYS TELL THE SAME 3 JOKES:

#1: THE WARM UP

#2: THE PSYCHIC

*ALWAYS GUESSED CHEERIOS

#3: THE TOUR DE FORCE

MY ATTEMPT:*

*MOMENT I DECIDED AGAINST PEDIATRICS

Things I Learned in Med School and Residency and Fellowship

MSI: If you're not in class, you should be studying

MSII: Nothing you learned in the last 2 years will ever be useful again

MSIII: You say you're tired?? Just you wait!

MSIV: This will probably be the best year of the rest of your life so enjoy it

PGY1: MEDICINE — You can eat dinner when I SAY you can EAT DINNER!

PGY2: PM&R — Five admits. I guess that means five rectal exams for you to do.

PGY3: STILL PM&R — I hope you like back pain, because every patient you're going to see has it

PGY4: PM&R AGAIN — You'll appreciate how easy you residents have it when you're an attending

FELLOWSHIP: RESEARCH — You can get started on your project right after you submit your 100-page IRB application and get it approved

ATTENDING: ?

(I can't wait...)

TIPS FOR GIVING A GOOD LECTURE

GIVE OUT PRIZES FOR RIGHT ANSWERS...

LIKE CANDY OR A CAR

USE A MICROPHONE TO AMPLIFY YOUR VOICE

BETTER YET, A *RAP* MICROPHONE

HAVE YOUR AUDIENCE SWITCH SEATS EVERY 10 MINUTES

AND EACH TIME, REMOVE ONE CHAIR

ONE WORD: 3D

($3 SURCHARGE WILL APPLY)

MED-SPEAK:

"TRY NOT TO BE ON CALL WITH JOHN. HE'S A TOTAL BLACK CLOUD!"

"I HATE MY SERVICE. I'VE GOT FIVE ROCKS."

"MR. SMITH HAD A 3 VESSEL CABBAGE YESTERDAY."

* BLACK CLOUD: BAD LUCK
ROCK: PATIENT WHO WILL NEVER EVER BE DISCHARGED
CABG: CORONARY ARTERY BYPASS GRAFT

TRUE STORIES OF PATIENT MIX-UPS

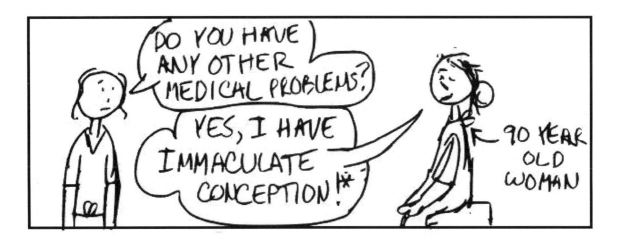

MED STUDENT Q&A:
WHAT IS A "DIRTY NETTER"?

a) AN ANATOMY ATLAS WHERE ALL THE ANATOMICAL IMAGES ARE OF PEOPLE HAVING SEX

b) AN ALCOHOLIC BEVERAGE CONTAINING GIN AND VERMOUTH

c) SEE PAGE 136 OF KAMA SUTRA

d) AN ATLAS COVERED IN LITTLE PIECES OF GUTS AND DRENCHED IN YOUR CADAVER'S BODILY JUICES

TIPS FOR EXAMINING ELDERLY PATIENTS

SPEAK LOUDLY

REDIRECT FREQUENTLY

EXPECT TO RESEMBLE AT LEAST ONE OF PATIENT'S GRANDCHILDREN

IN ORDER TO BUILD RAPPORT, ASK ABOUT BOWELS

MINI-MENTAL STATUS EXAM
(AS ADMINISTERED TO POST-CALL RESIDENTS)

COUNT BACKWARDS FROM 100 BY 7's

READ THE SENTENCE AND DO WHAT IT SAYS

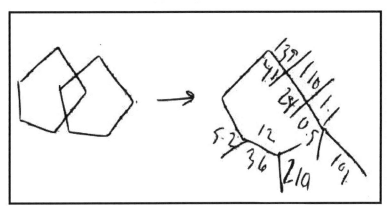

COPY THE DRAWING

THE MINI MENTAL STATUS EXAM.
"REMEMBER THE FOLLOWING 3 WORDS: APPLE, TABLE, PENNY"
RESPONSES YOU DON'T WANT TO HEAR:

MENTAL STATUS ASSESSMENT
MOST COMMON ANSWERS MY PATIENTS GIVE TO THE QUESTION: "WHO IS THE PRESIDENT?"

MOST COMMON ANSWER TO: "WHO WAS THE PRESIDENT BEFORE OBAMA?"

MEDICAL ABBREVIATIONS: PDA

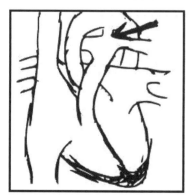
PATENT DUCTUS ARTERIOSUS

PERSONAL DIGITAL ASSISTANT

PUBLIC DISPLAY OF AFFECTION

THE 6 GRADES OF HEART MURMURS

GRADE I:
CAN ONLY BE HEARD WITH SPECIAL CARDIOLOGIST STETHOSCOPE

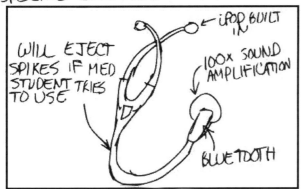

GRADE II:
LOUD ENOUGH FOR MED STUDENT TO PRETEND TO HEAR

GRADE III:
SLEEPY RESIDENT CAN PICK UP

GRADE IV:
STETHOSCOPE BOUNCES ON CHEST

GRADE V:
MED STUDENT CAN ALMOST HEAR

GRADE VI:
PATIENT'S ROOMMATE WILL COMPLAIN

TERRIBLE BABY NAMES:

MELENA
MRSA
LUPUS
VARICELLA
JCAHO
PLACENTA
CHLAMYDIA
GONORRHEA
NEOGENESIS
AIDEN
FISTULA

"I LOVE YOU, LITTLE GUAIAC"

MOST COMMON PASSWORDS USED BY DOCTORS AND MED STUDENTS

"YOUR AASSWORD IS "PASSWORD"? THAT'S NOT VERY SECURE"

"OH NO. SOMEONE MIGHT CHECK MY PATIENTS' SODIUM FOR ME OR SOMETHING"

MONEY
DEBT
MASSIVEDEBT
VALSALVA
MEDICINESUX
PASSWORD
FUCKYOU
FUCKME
FUCKME69!
12345
12345678

8675309
GOMER
TURF
NCC1701*
DOCTOR
DRHOUSE
BOOBZ
CREMASTER
CPSOB
PORSCHE

*MOST COMMON PASSWORD OF FUTURE NEUROLOGISTS

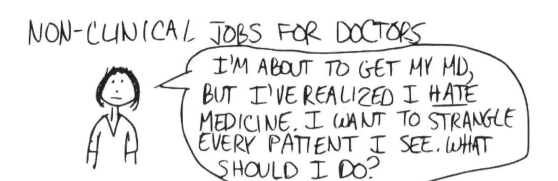

NON-CLINICAL JOBS FOR DOCTORS

"I'M ABOUT TO GET MY MD, BUT I'VE REALIZED I HATE MEDICINE. I WANT TO STRANGLE EVERY PATIENT I SEE. WHAT SHOULD I DO?"

SOME OPTIONS:

DRUG COMPANY REP

FINANCE/CONSULTING

STAY AT HOME MOM

CASHIER AT STARBUCKS

WHERE DID YOU LOSE YOUR PAGER?

IN TOILET

"ACCIDENTALLY" HURLED OUT WINDOW

LEFT INSIDE PATIENT

WHAT IS WRONG WITH THIS MEDICAL STUDENT'S EYES?

a) THEY ARE OPEN
b) THEY ARE EVIL
c) BITEMPORAL FRUITOPIA

WHAT TO DO WHEN YOU SEE AN ER PATIENT WHO STUCK A PENCIL IN HIS RECTUM:

STEP 1: BE PROFESSIONAL

STEP 2: REVIEW IMAGING

STEP 3: EMBELLISH ON INTERNET

MEDICAL CONDITIONS MY HUSBAND THINKS EXIST (BUT DON'T):

"ISN'T THERE SOME CONDITION WHERE YOU HAVE LOTS AND LOTS OF TINY PENISES? LIKE, HUNDREDS?"

CENTIPENIS: NOT AS HANDY AS YOU'D THINK

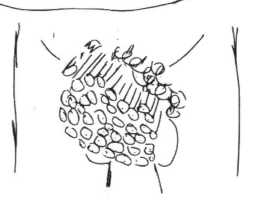

THE 10 KINDS OF PRIMARY CARE PATIENTS

THE MYSTERY COMPLAINT
MULTIPLE MEDICAL PROBLEMS (MMP)

PISSED OFF
NO HABLA INGLES
GRATEFUL

DRUG SEEKER
Ψ
CHRONIC BACK PAIN

MULTIPLE COMPLAINTS
PATIENT + ADULT CHILDREN

EVOLUTION OF THE PHYSICAL EXAM

THE ER

THE MED STUDENT

THE RESIDENT

THE ATTENDING

CONSULT SERVICE

DISCHARGE SUMMARY, ONE WEEK LATER:

Cardiac: Reg rate and rhythm, no murmurs, rubs, or gallops, purple monkey dishwasher.

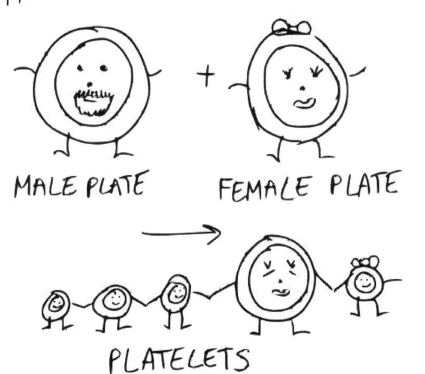

MEDICAL MISSPELLINGS

HIPPA

PUSS

PMNR

PERONEAL/PERINEAL
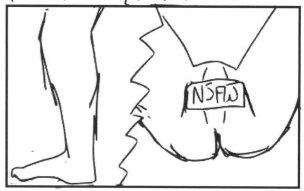

OPHTHALMOLOGY
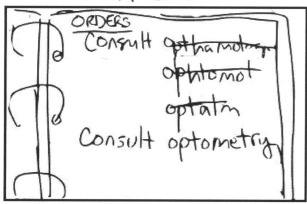

HOW TO EXAMINE A TESTICULAR MASS

1: INTRODUCE YOURSELF

2: TRY NOT TO SAY SOMETHING SUGGESTIVE WHILE FEELING TESTES

3: TURN OFF LIGHT IN EXAM ROOM AND TRANSILLUMINATE TESTICLE POSTERIORLY

4: BLOW RAPE WHISTLE
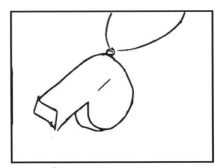

WAYS TO BREAK STERILE IN THE OR:

SCRATCHING YOUR NOSE

TOUCHING STERILE GOWN WITH SCRUBBED UNGLOVED HAND

BREATHING WRONG

WAYS NOT TO BREAK STERILE

BEING ELBOW-DEEP IN SOMEONE'S BOWELS

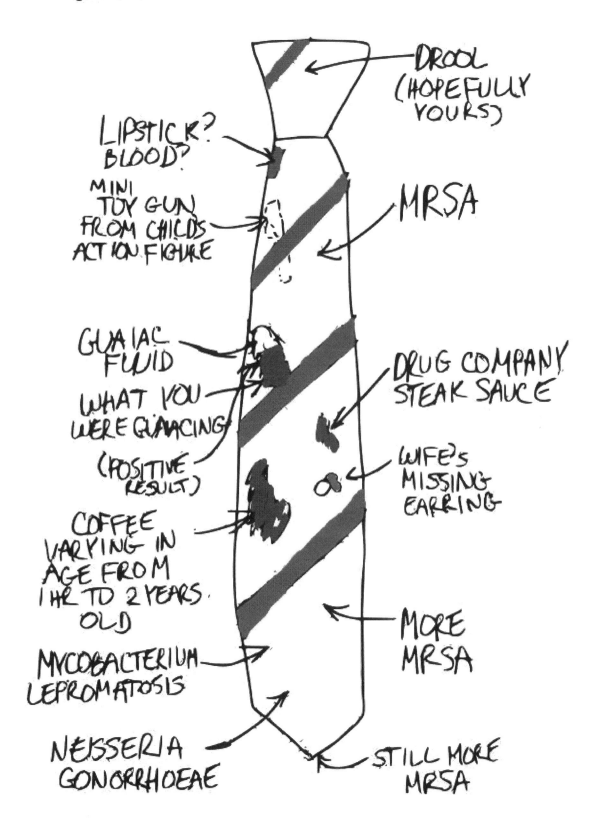

FUN WITH THE HOSPITAL OPERATORS

WAYS PATIENTS HAVE TOLD ME I LOOK YOUNG

HEART CONDITIONS YOU DON'T LEARN TO TREAT IN MED SCHOOL

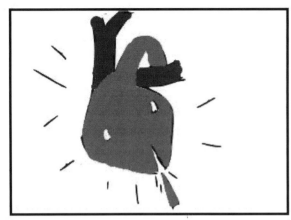

HEART OF GLASS

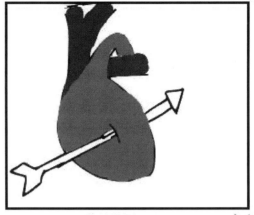

SHOT THROUGH THE HEART

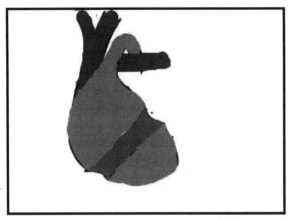

GROOVE IS IN THE HEART

TOTAL ECLIPSE OF THE HEART

THE 7 TYPES OF PHYSICIAN HANDWRITING

5 YEAR OLD HANDWRITING:

Patient seen and examined by nees tnight

IMMACULATE, ILLEGIBLE SCRIPT:

Abdomen soft nontender, nondistended

SANSKRIT:

[Sanskrit-like script]

EVERY 4TH WORD LEGIBLE:

critical STAT!

EVERY WORD MUST TOUCH LINE MARGINS:

Patient is alert and oriented x3

TEENY TINY:

Patient has history of hypertension and diabetes

HAD 30 SECONDS TO WRITE NOTE:

[squiggle]

Made in the USA
Lexington, KY
30 October 2013